Opportunity class style test-Revision Part-1

This study material has been produced by using the resource available in © State of New South Wales (Department of Education), 2023 as a guide. The author does not have any connection with © State of New South Wales (Department of Education) and this work has not been endorsed by © State of New South Wales (Department of Education).

Author has taken reasonable effort to make this work free from errors and mistakes however there is no guarantee that this material is free from errors or mistakes.

Opportunity class style test-Revision Part-1

Clockwise and anticlockwise

As shown below a G-shape is made from squares, Question 1 to13 is based on the G shape below.

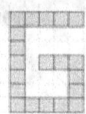

1. Which of the below will show how the G-shape will look after it is turned half of a turn anti-clockwise?

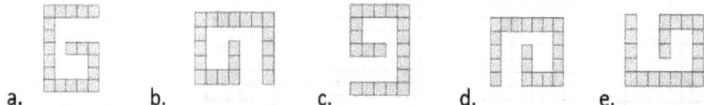

a.   b.   c.   d.   e.

2. Which of the below will show how the G-shape will look after it is turned quarter of a turn anti-clockwise

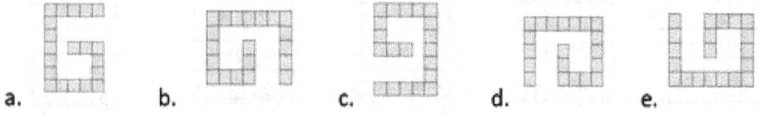

a.   b.   c.   d.   e.

3. Which of the below will show how the G-shape will look after it is turned three quarters of a turn anti clockwise

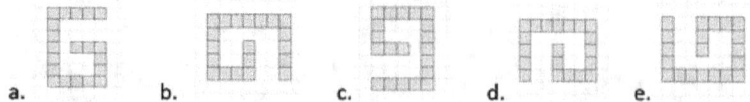

a.   b.   c.   d.   e.

Opportunity class style test-Revision Part-1

4. Which of the below will show how the G-shape will look after it is turned 90 degrees anti clockwise?

5. Which of the below will show how the G-shape will look after it is turned 45 degrees anti clockwise?

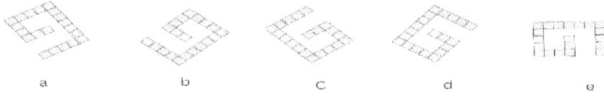

6. Which of the below will show how the G-shape will look after it is turned 180 degrees anti clockwise?

7. Which of the below will show how the G-shape will look after it is turned half of a turn clockwise?

8. Which of the below will show how the G-shape will look after it is turned quarter of a turn clockwise?

9. Which of the below will show how the G-shape will look after it is turned three quarters of a turn clockwise?

Opportunity class style test-Revision Part-1

10. Which of the below will show how the G-shape will look after it is turned 180 degrees clockwise?

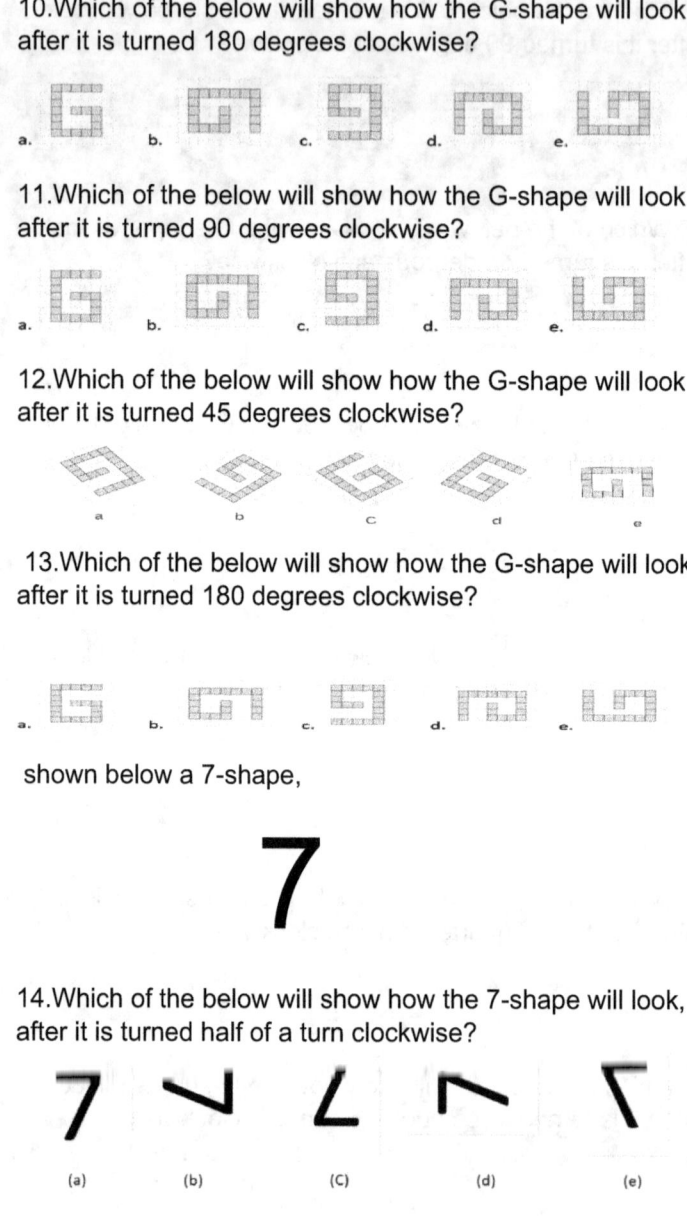

11. Which of the below will show how the G-shape will look after it is turned 90 degrees clockwise?

12. Which of the below will show how the G-shape will look after it is turned 45 degrees clockwise?

13. Which of the below will show how the G-shape will look after it is turned 180 degrees clockwise?

shown below a 7-shape,

14. Which of the below will show how the 7-shape will look, after it is turned half of a turn clockwise?

Opportunity class style test-Revision Part-1

15. Which of the below will show how the 7-shape will look after it is turned quarter of a turn clockwise?

16. Which of the below will show how the 7-shape will look after it is turned three quarters of a turn clockwise?

17. Which of the below will show how the 7-shape will look after it is turned 180 degrees clockwise?

As shown below a 9-shape,

18. Which of the below will show how the 9-shape will look after it is turned half of a turn clockwise?

19. Which of the below will show how the 9-shape will look after it is turned quarter of a turn clockwise?

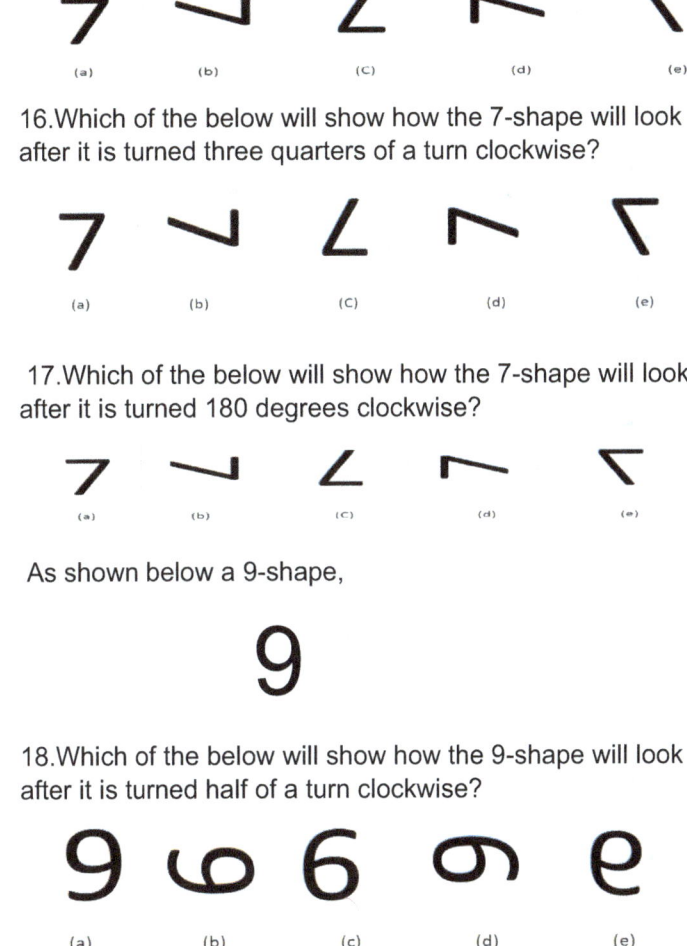

Opportunity class style test-Revision Part-1

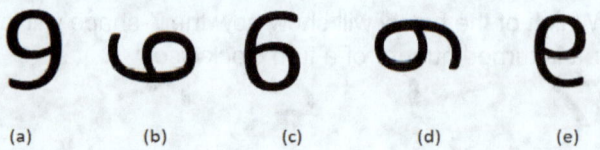

20. Which of the below will show how the 9-shape will look after it is turned three quarters of a turn clockwise?

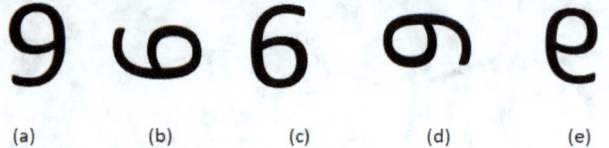

21. Which of the below will show how the 9-shape will look after it is turned 180 degrees clockwise?

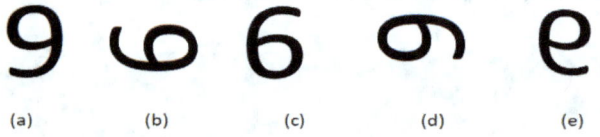

22. Which of the below will show how the upward pointed arrow shape will look when

22. after it is turned a half of a turn clockwise?

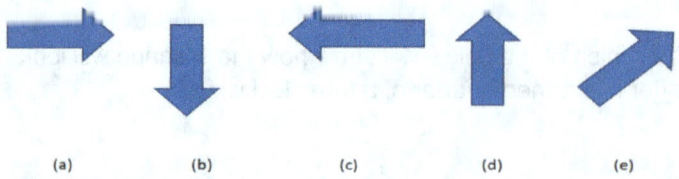

6

Opportunity class style test-Revision Part-1

23. Which of the below will show how the upward pointing shape will look after it is turned quarter of a turn clockwise?

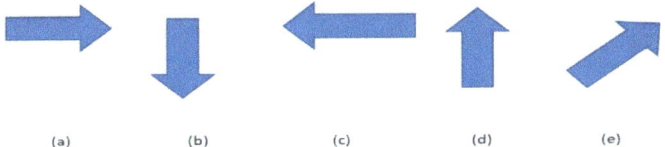

(a)   (b)   (c)   (d)   (e)

24. Which of the below will show how the upward pointing arrow shape will look after it is turned three quarters of a turn clockwise?

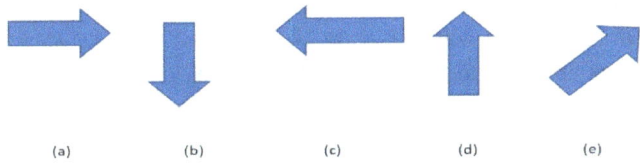

(a)   (b)   (c)   (d)   (e)

25. Which of the below will show how the upward pointing arrow shape will look after it is turned 180 degrees clockwise?

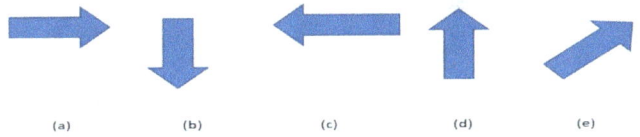

(a)   (b)   (c)   (d)   (e)

Opportunity class style test-Revision Part-1

As shown below a Q-shape,

26. Which of the below will show how the Q-shape will look after it is turned half of a turn clockwise?

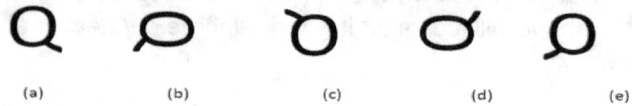

(a)　　　(b)　　　(c)　　　(d)　　　(e)

27. Which of the below will show how the Q-shape will look after it is turned quarter of a turn clockwise?

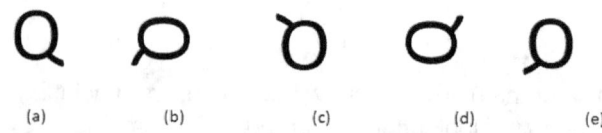

(a)　　　(b)　　　(c)　　　(d)　　　(e)

28. Which of the below will show how the Q-shape will look after it is turned three quarters of a turn clockwise?

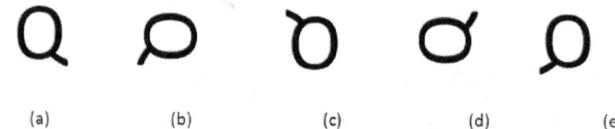

(a)　　　(b)　　　(c)　　　(d)　　　(e)

29. Which of the below will show how the Q-shape will look after it is turned quarters of a turn anti clockwise?

(a)　　　(b)　　　(c)　　　(d)　　　(e)

Opportunity class style test-Revision Part-1

30. Which of the below will show how the Q-shape will look after it is turned 180 degrees clockwise?

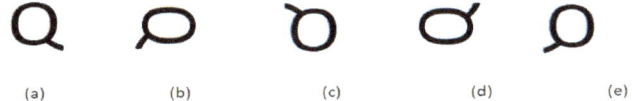

(a)　　(b)　　(c)　　(d)　　(e)

As shown below a upward facing smiley-shape,

31. Which of the below will show how the upward facing smiley-shape will look after it is turned half of a turn clockwise?

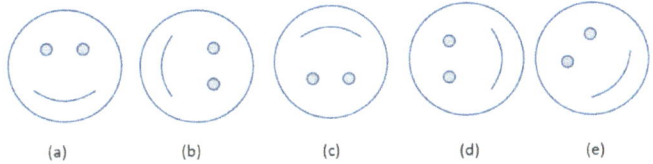

(a)　　(b)　　(c)　　(d)　　(e)

32. Which of the below will show how the upward facing smiley shape will look after it is turned quarter of a turn clockwise?

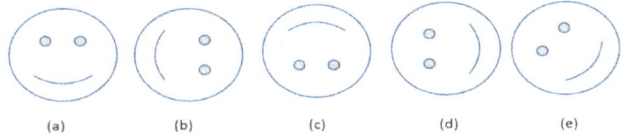

(a)　　(b)　　(c)　　(d)　　(e)

33. Which of the below will show how the upward facing smiley-shape will look after it is turned three quarters of a turn clockwise?

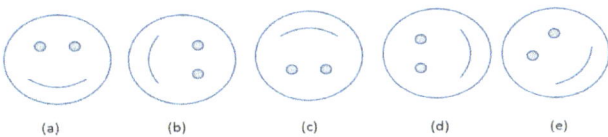

(a)　　(b)　　(c)　　(d)　　(e)

34. Which of the below will show how the upward facing smiley-shape will look after it is turned quarter of a turn clockwise?

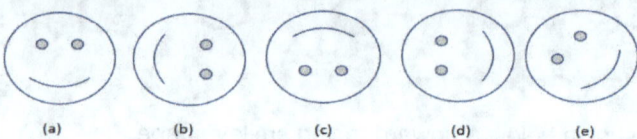

35. Which of the below will show how the upward facing smiley-shape will look after it is turned 180 degrees clockwise?

36. The ball below makes a quarter of a turn in clockwise every time it jumps on the ground.

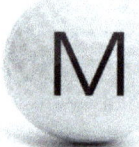

The ball was dropped on the ground as per below and the letter M was upright. In 5 jumps how will the ball appear?

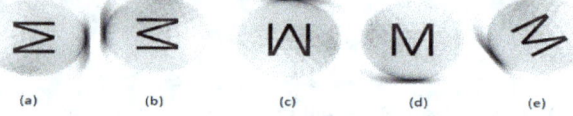

37. The ball below makes a quarter of a turn in anti-clockwise every time it jumps on the ground.

The ball was dropped on the ground as per below and the letter M was upright. In 5 jumps how will the ball appear?

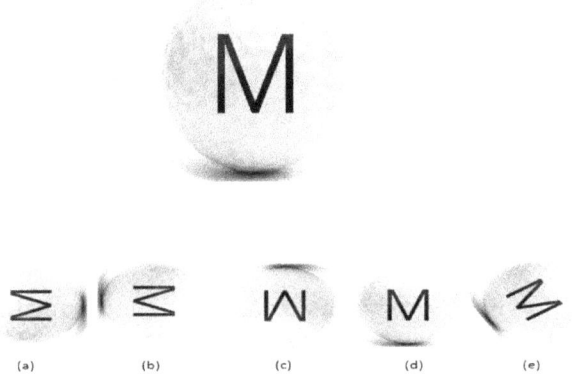

(a) (b) (c) (d) (e)

38. The ball below makes a quarter of a turn in anti-clockwise every time it jumps on the ground.

The ball was dropped on the ground as per below and the letter M was upright. In 3 jumps how will the ball appear?

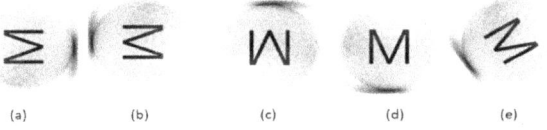

(a) (b) (c) (d) (e)

39. The ball below makes a quarter of a turn in anti-clockwise every time it jumps on the ground.

The ball was dropped on the ground as per below and the letter M was upright. In 11 jumps how will the ball appear?

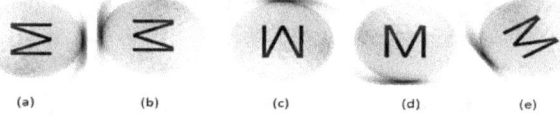

40. The ball below makes a quarter of a turn in anti-clockwise every time it jumps on the ground.

The ball was dropped on the ground as per below and the letter M was upright. In 10 jumps how will the ball appear?

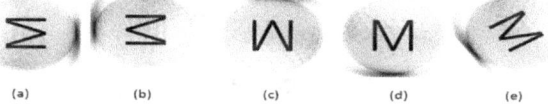

41. The ball below makes a quarter of a turn in clockwise every time it jumps on the ground.

The ball was dropped on the ground as per below and the letter M was upright. In 7 jumps how will the ball appear?

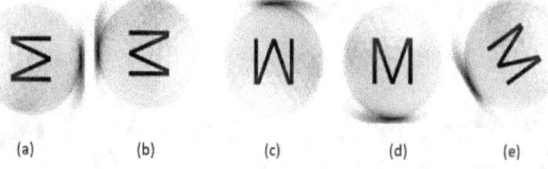

42.The ball below makes 3 quarters of a turn in clockwise every time it it jumps on the ground.

The ball was dropped on the ground as per below and the letter M was upright. In 3 jumps how will the ball appear?

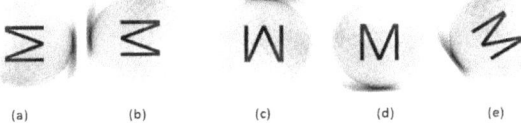

(a)　　(b)　　(c)　　(d)　　(e)

43.The ball below makes 5 quarters of a turn in anti-clockwise every time it jumps on the ground.

The ball was dropped on the ground as per below and the letter M was upright. In 2 jumps how will the ball appear?

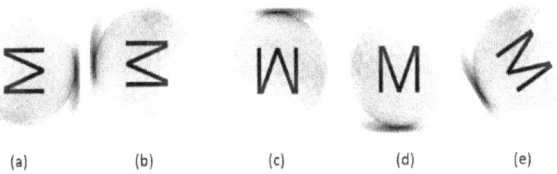

(a)　　(b)　　(c)　　(d)　　(e)

44. The ball below makes a half of a turn in clockwise every time it jumps on the ground.

The ball was dropped on the ground as per below and the letter M was upright. In 3 jumps how will the ball appear?

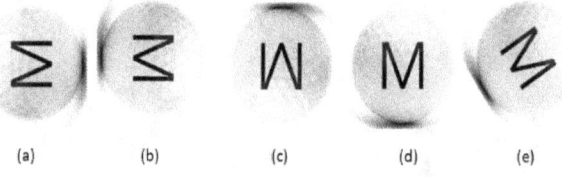

45. The ball below makes a full turn in clockwise every time it jumps on the ground.

The ball was dropped on the ground as per below and the letter M was upright. In 3 jumps how will the ball appear?

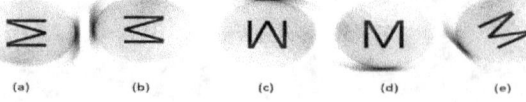

46. The ball below jumps 1 metre in the first time it collides with the ground. From next time it touches it doubles the distance of the previous jump. In the 5th jump what would be the height of the jump?

(a) 1m
(b) 2m
(c) 4m
(d) 8m
(e) 16m

47. The scale below is not balanced

Which of the below will balance the scale

1. add 6 Kgs to the left and 2 Kgs to the right
2. add 4 Kgs to the left and 2 Kgs to the right
3. add 4 Kgs to the left only
4. add 6 Kgs to the right only

A. method 1 only

B. method 2 only

C. method 3 only

D. methods 1 and 2 only

E. methods 1 and 3 only

Opportunity class style test-Revision Part-1

48. To balance the scale below

Which of the below will balance the scale
1. add 6.5 Kgs to the left and 3.25 Kgs to the right
2. add 3.25 Kgs to the left and 6.5 Kgs to the right
3. add 3.25 Kgs to the left only
4. add 3.25 Kgs to the right only

A. method 1 only

B. method 2 only

C. method 3 only

D. methods 1 and 2 only

E. methods 2 and 4 only

49. To balance the scale below

Which of the below will balance the scale

1. add 4 Kgs to the left and 8 Kgs to the right
2. add 8 Kgs to the left and 4 Kgs to the right
3. add 4 Kgs to the left only
4. add 4 Kgs to the right only

A. method 1 only

B. method 2 only

C. method 3 only

D. methods 1 and 2 only

E. methods 2 and 3 only

50. To balance the scale below

Which of the below will balance the scale

1. add   2 Kgs to the left and 2.09 Kgs to the right
2. add   2 Kgs to the left and 2.9 Kgs to the right
3. add   0.09 Kgs to the left only
4. add   0.09 Kgs to the right only

A. method 1 only

B. method 2 only

C. method 3 only

D. methods 1 and 2 only

E. methods 1 and 4 only

51. To balance the scale below

Which of the below will balance the scale

1. add 3.1Kgs to the left and 6,2 Kgs to the right
2. add 6.2 Kgs to the left and 3.1Kgs to the right
3. add 3.1Kgs to the left only
4. add 3.1 Kgs to the right only

A. method 1 only

B. method 2 only

C. method 3 only

D. methods 1 and 2 only

E. methods 1 and 4 only

52. To balance the scale below

Which of the below will balance the scale

1. add 0.4 Kgs to the left and 0.2 Kgs to the right
2. add 0.2 Kgs to the left and 0.4 Kgs to the right
3. add 1.4 Kgs to the left and 0.2 Kgs to the right
4. add 0.2 Kgs to the left only
5. add 0.2 Kgs to the right only

A. method 1 only

B. method 2 only

C. method 3 only

D. methods 1 and 2 only

E. methods 2 and 5 only

53. Ruler is broken. scale represents 1 cm.

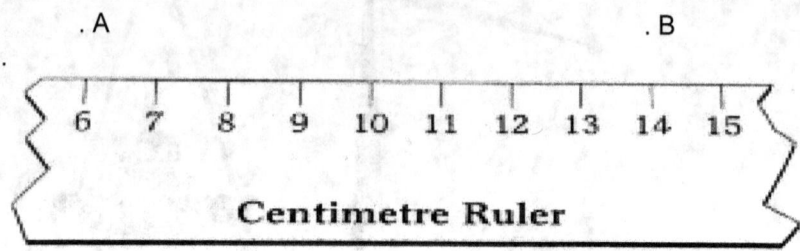

What is the correct length between point A and point B in cm's.

A 8 cm

B 14cm

C 4 cm

D 140 cm

E 1400 cm

54. The below diagram show the distance between 2 trees. Mango tree is located northwest to the Coconut tree. Sam uses a broken ruler to measure the distance between the 2 trees. the ruler measures in Centimetres.
1 centimetre represents 5 meters in the real distance.

What is the correct distance between 2 trees in meters?

A 8 meters

B 40 meters

C 4 meters

D 0.08 meters

E 80 meters

55. The below diagram show the distance between 2 trees. Mango tree is located southwest to the Coconut tree. Sam uses a broken ruler to measure the distance between the 2 trees. the ruler measures in Centimetres. 1 centimetre represents 7 meters in the real distance.

What is the correct distance between 2 trees in meters?

A 49 meters

B 63 meters

C 7 meter

D 56 meters

E 8 meters

56. The below diagram show the distance between 2 trees. Mango tree is located Southwest to the Coconut tree. Sam uses a broken ruler to measure the distance between the 2 trees. the ruler measures in Centimetres. 1 centimetre represents 2 meters in the real distance.

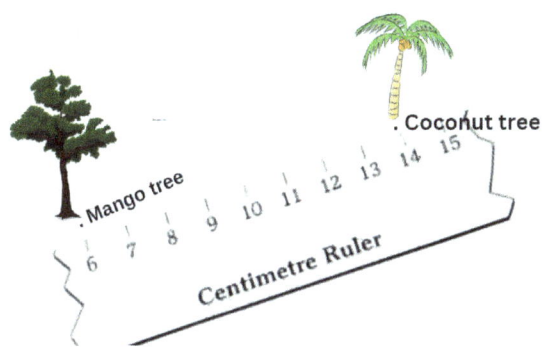

What is the correct distance between 2 trees in meters?

A 16 meters

B 8 meters

C 160 meters

D 80 meters

E 0.8 meters

57. The below diagram show the distance between 2 trees. Mango tree is located Southwest to the Coconut tree. Sam uses a broken ruler to measure the distance between the 2 trees. the ruler measures in Centimetres. 1 centimetre represents 17 meters in the real distance.

What is the correct distance between 2 trees in meters?

A 13.6 meters

B 136 meters

C 1.36 meter

D 80 meters

E 56 meters

Opportunity class style test-Revision Part-1

58. The below diagram show the distance between 2 trees. Mango tree is located Southwest to the Coconut tree. Sam uses a broken ruler to measure the distance between the 2 trees. the ruler measures in Centimetres.

1 millimetre represents 12 meters in the real distance.

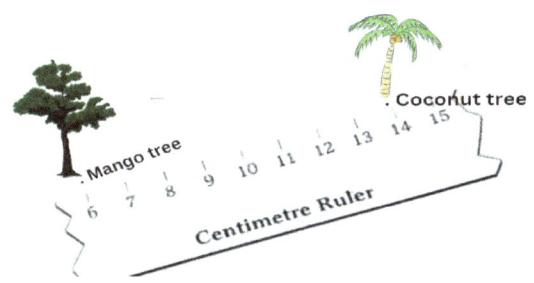

What is the correct distance between 2 trees in meters?

A 96 meters
B 100 meters
C 9.6 meter
D 960 meters
E 120 meters

59. The below diagram show the distance between 2 trees. Mango tree is located Southwest to the Coconut tree. Sam uses a broken ruler to measure the distance between the 2 trees. the ruler measures in Centimetres. 5 millimetre represents 2 meters in the real distance.

What is the correct distance between 2 trees in meters?

A 10 meters

B 320 meters

C 32 meters

D 3.2 meters

E 16 meters

60. The below diagram show the distance between 2 trees. Mango tree is located Southwest to the Coconut tree. Sam uses a broken ruler to measure the distance between the 2 trees. the ruler measures in Centimetres. 1 millimetre represents half of a meters in the real distance.

What is the correct distance between 2 trees in meters?

A 10 meters

B 40 meters

C 4 meters

D 400 meters

E 0.4 meters

61. The below diagram show the distance between 2 trees. Mango tree is located Southwest to the Coconut tree. Sam uses a broken ruler to measure the distance between the 2 trees. the ruler measures in Centimetres. 1 Centimetre represents quarter of a meters in the real distance.

What is the correct distance between 2 trees in meters?

A 80 meters

B 20 meters

C 40 meters

D 2 meters

E 4 meters

62. The below diagram show the distance between 2 trees. Mango tree is located Southwest to the Coconut tree. Sam uses a broken ruler to measure the distance between the 2 trees. the ruler measures in Centimetres. 1 millimetre represents three quarters of a meter in the real distance.

What is the correct distance between 2 trees in meters?

A 6 meters

B 60 meters

C 600 meter

D 20 meters

E 40 meters

63. The below diagram show the distance between 2 trees. Mango tree is located Southwest to the Coconut tree. Sam uses a broken ruler to measure the distance between the 2 trees. the ruler measures in Centimetres. one tenth of a centimetre represents 1 metre in the real distance.

What is the correct distance between 2 trees in meters?

A 80 meters

B 800 meters

C 8 meters

D 40 meters

E 4 meters

64. The below diagram show the distance between 2 trees. Mango tree is located Southwest to the Coconut tree. Sam uses a broken ruler to measure the distance between the 2 trees. the ruler measures in Centimetres. one hundredth of a centimetre represents 1 metre in the real distance.

What is the correct distance between 2 trees in meters?

A 80 meters

B 800 meters

C 8 meter

D 8000 meters

E 40 meters

65. When the puzzle below is completed, each row and each column will contain a correct number sentence.

What number does letter Z represent?

4	×	6	=	24
×				÷
3				6
=				=
Q	÷	B	=	Y
÷				×
6				9
=				=
R	×	18	=	Z

A 36
B 9
C 12
D 4
E 3

66. When the puzzle below is completed, each row and each column will contain a correct number sentence.

What number does letter Y represent?

4	×	6	=	24
×				÷
3				6
=				=
Q	÷	B	=	Y
÷				×
6				9
=				=
R	×	18	=	Z

- A 36
- B 9
- C 12
- D 4
- E 3

67. When the puzzle below is completed, each row and each column will contain a correct number sentence.

What number does letter B represent?

4	×	6	=	24
×				÷
3				6
=				=
Q	÷	B	=	Y
÷				×
6				9
=				=
R	×	18	=	Z

A   36
B   9
C   12
D   4
E   3

68. When the puzzle below is completed, each row and each column will contain a correct number sentence. What number does letter R represent?

4	×	6	=	24
×				÷
3				6
=				=
Q	÷	B	=	Y
÷				×
6				9
=				=
R	×	18	=	Z

A 36
B 9
C 2
D 4
E 3

Opportunity class style test-Revision Part-1

69. When the puzzle below is completed, each row and each column will contain a correct number sentence.

What sign does letter N represent?

4	x	6	=	24
x				N
3				6
=				=
Q	÷	B	=	4
÷				x
6				9
=				=
R	x	18	=	Z

A +
B X
C =
D -
E ÷

38

70. When the puzzle below is completed, each row and each column will contain a correct number sentence.

What sign does letter N represent?

4	×	6	=	24
×				÷
3				6
=				=
Q	÷	B	=	Y
÷				×
6				9
=				=
R	**N**	18	=	Z

A +
B ×
C =
D -
E ÷

71. When the puzzle below is completed, each row and each column will contain a correct number sentence.

What sign does letter N represent?

	7			147
	×			N
	14			3
	=			=
		÷	2	=

A +
B ×
C =
D -
E ÷

Opportunity class style test-Revision Part-1

72. When the puzzle below is completed, each row and each column will contain a correct number sentence.

What number does letter N represent?

4				N
x				=
8				55
=				+
-				=
7				9
=				x
	÷	5	=	

A 110

B 100

C 45

D 32

E 64

Opportunity class style test-Revision Part-1

73. When the puzzle below is completed, each row and each column will contain a correct number sentence.

What sign does letter N represent?

7				109
x				=
8				55
=				N
-				=
14				9
=				x
	÷	7	=	

A   +
B   -
C   x
D   =
E   ÷

Opportunity class style test-Revision Part-1

74. When the puzzle below is completed, each row and each column will contain a correct number sentence.

What sign does letter N represent?

7			-	9	=	100
×		=				=
8		55				5
=		+				×
-		=				×
14		9				1
=		N				×
	÷	7	=			5

A +

B -

C ×

D =

E ÷

75. Which one of the following numbers is closest to 2?

A 2 ½

B 1 ¾

C 1 ⅔

D 2 □

E 2 ⅛

43

76. Which one of the following numbers is closest to 1?

A 1 ½
B ¾
C ⅔
D 1 ☐
E 1 ⅛

77. What are the correct answers for 2 - 2/8 from below?

1. 1 ¾
2. 1 6/8
3. 1 12/16
4. 1 14/16
5. 1 7/8

A 1 only
B 2 only
C 3 only
D 1 & 2 only
E 1, 2 & 3 only

78. What are the correct answers for 2- ½ from below?

1. 1 2/4
2. 1 ½
3 1 8/16
4 1 3/6
5 1 5/10

A 1 only
B 2 only
C 3 only
D 1 &2 only
E all

79. What are the correct answers for 2- 3/6 from below?

1. 1 2/4
2. 1 1/2
3 1 8/16
4 1 3/6
5 1 5/10

A 1 only
B 2 only
C 3 only
D 1 &2 only
E all

Opportunity class style test-Revision Part-1

80. What are the correct answers for 2- 9/18 from below?
1. 1 2/4
2. 1 1/2
3 1 8/16
4 1 3/6
5 1 5/10

A 1 only
B 2 only
C 3 only
D 1 &2 only
E all

81. What are the correct answers for 2- 33 / 99 from below?
1. 1 2/3
2. 1 4/6
3 1 6/9
4 1 10/15
5 1 7/21

A 1 only
B 2 only
C 3 only
D 1 &2 only
E all

Opportunity class style test-Revision Part-1

82. What are the correct answers for 2- 11/99 from below?

1. 1 8/9
2. 1 4/9
3. 1 6/9
4. 1 10/15
5. 1 7/21

A 1 only
B 2 only
C 3 only
D 1 &2 only
E all

83. What are the correct answers for 2 + 11/99 from below?

1. 1 8/9
2. 2 1/9
3. 1 89/99
4. 2 10/99
5. 1 11/9

A 1 only
B 2 only
C 3 only
D 1 &2 only
E all

84. What is the total of 9 thousands, 9 hundreds, 9 tens and 9 ones?

A 9999

B 99999

C 999

D 99099

E 9909

85. What is the total of 12 thousands, 12 hundreds, 12 tens and 12 ones?

A 13332

B 13320

C 12222

D 13333

E 12332

86. What is the total of 13 thousands, 13 hundreds, 13 tens and 13 ones?

A 14443

B 144430

C 13333

D 14333

E 14444

87. What is the total of 14 thousands, 14 hundreds, 14 tens and 14 ones?

A 15554

B 15555

C 15550

D 14554

E 14444

88. What is the total of 15 thousands, 15 hundreds, 15 tens and 15 ones?

A 16665

B 15555

C 15550

D 16666

E 16565

89. What is the total of 16 thousands, 16 hundreds, 16 tens and 16 ones?

A 17776

B 15555

C 16666

D 17777

E 17767

90. What is the total of 17 thousands, 17 hundreds, 17 tens and 17 ones?

A 18887

B 11777

C 18878

D 18888

E 18787

91. What is the total of 18 thousands, 18 hundreds, 18 tens and 18 ones?

A 19998

B 19999

C 18898

D 18988

E 19989

92. What is the total of 19 thousands, 19 hundreds, 19 tens and 19 ones?

A 21109

B 19999

C 21119

D 21209

E 21199

93. What is the total of 20 thousands, 20 hundreds, 20 tens and 20 ones?

A 22220

B 20220

C 22020

D 20200

E 22222

94. What is the total of 49 thousands, 49 hundreds, 49 tens and 49 ones?

A 54439

B 53339

C 49499

D 54039

E 54339

Opportunity class style test-Revision Part-1

95. What are the rounded prices for the below to the nearest dollar?

$ 6.59 $ 6.49 $ 8.95

A $6.00, $6.00, $9.00
B $7.00, $6.00, $9.00
C $7.00, $7.00, $9.00
D $6.00, $7.00, $9.00
E $7.00, $7.00, $8.00

96. Cupid's mum gave him $30 to spend on the 3 items below

To estimate the total cost, Cupid rounds each price to the nearest dollar and then adds the rounded values together.

$ 6.59 $ 6.49 $ 8.95

What would be the estimated change
A $8.00
B $7.00
C $10.00
D $6.00
E $9.00

97. Cupid's mum gave him $55 to spend on the 3 items below

To estimate the total cost, Cupid rounds each price to the nearest dollar and then adds the rounded values together.

| $ 6.59 | $ 6.49 | $ 8.95 |

What would be the estimated change

A $8.00

B $7.00

C $33.00

D $6.00

E $9.00

98. Cupid's mum gave him one $5 note two $10 note and one $20 note to buy some items. At school he lost one note and left with only 3 notes. He bought the 3 items below and got a change of $12.97

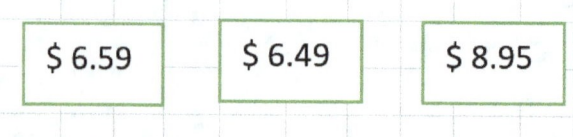

What is the note he lost

A $5

B $10

C $20

D none

E all

Opportunity class style test-Revision Part-1

99. Cupid's mum gave him one $5 note two $10 note and one $20 note to buy some items. At school he lost two notes and left with only 2 notes. He bought the 3 items below and got a change of $7.97

| $ 6.59 | $ 6.49 | $ 8.95 |

What are the notes he lost

A $5

B One $10 and $5

C $20

D Both $10 notes

E $5 and $20

100. Pilki has got $10 from his Mum to buy the breakfast from school canteen. He decided to buy each of the below. To estimate the change he gets he decided to round the values to the nearest dollar.

Sandwitch	$2.97
Muffin	$2.45
Banana	$1.15

How much change will Pilki get when rounded to the nearest dollar?

A $4

B $3

C $2

D $5

E $6

101. Ash has the 2 objects that is comprised of 20 cubes each. How many objects with 4 cubes can he make?

A 4

B 5

C 10

D 20

E 8

102. Beau is building a fence using bricks, one layer at a time. Rule is he has to always complete one layer before he starts the next layer. Each layer is made from 7 bricks. Beau has 36 bricks and he uses them all. Which layer is she working on when she lays her last brick?

A 4th

B 5th

C 6th

D 7th

E 8th

103. Beau is building a fence using bricks, one layer at a time. Rule is he has to always complete one layer before he starts the next layer. As per the plan each layer is made from 8 bricks and Beau purchased 32 bricks to do the job. The fence has to be completed with 4 layers. How ever he accidentally used 9 bricks for each layer and she uses them all. How many more bricks does need to complete the job.

A 4

B 5

C 3

D 2

E 6

104. Beau is building a fence using bricks, one layer at a time. Rule is he has to always complete one layer before he starts the next layer. As per the plan each layer is made from 8 bricks and Beau purchased 40 bricks to do the job. When he completed 4 layers he decided to continue rest of the layers with 4 bricks in each layer . How many extra layers he will end up once the job is compete than the original plan.

A 4

B 5

C 3

D 2

E 1

105. There is a box of 26 coloured discs.

- 10 discs are red.
- 2 discs are green.
- 13 discs are blue.
- 1 disc is yellow.

John picks one disc at random from the box. Which of these statements are correct?

1 It is certain that he picks a blue disc.

2 It is equally likely that he picks a blue disc or does not pick a blue disc.

3 It is possible for him to pick a yellow disc.

A none of them

B statements 1 and 2 only

C statements 1 and 3 only

D statements 2 and 3 only

E statements 1, 2 and 3

106. There is a box of 28 coloured discs.

- 10 discs are red.
- 3 discs are green.
- 14 discs are blue.
- 1 disc is yellow.

John picks one disc at random from the box. Which of these statements are correct?

1 It is certain that he picks a blue disc.

2 It is equally likely that he picks a blue disc or does not pick a blue disc.

3 It is possible for him to pick a black disc.

4 It is more likely to select a blue disk

A none of them

B statements 1 and 2 only

C statements 1 and 3 only

D statements 2 and 4 only

E statements 1, 2 and 3

107. There is a box of 99 coloured discs.

- 30 discs are red.
- 22 discs are green.
- 33 discs are blue.
- 14 discs are yellow.

Gooey picks one disc at random from the box. Which of these statements are correct?

1 It is certain that he picks a blue disc.

2 higher likeliness to pick a blue disc than any other single colour disc

3 It is possible for him to pick a yellow disc

4 It is equally likely to pick a blue disk or not

A statements 1 and 4 only

B statements 1 and 2 only

C statements 1 and 3 only

D statements 2 and 3 only

E statements 1, 2 and 3

108. There is a box of 10 coloured discs.

- 2 discs are red.
- 3 discs are green.
- 1 discs are blue.
- 4 disc is yellow.

Toto picks one disc at random from the box. Which of these statements are correct?

1 It is certain that he picks a blue disc.

2 least likeliness to pick a blue disc

3 It is possible for him to pick a yellow disc.

4 It is equally likely to pick one of the colours from red and green or one from the colour blue and yellow discs

5 It is possible to pick a black disc

A statements 2,3 and 4 only

B statements 1 and 2 only

C statements 1 and 3 only

D statements 2 and 3 only

E statements 1, 2,3 and 5

109. Shaun has a 0.5 litre measuring jug. It contains 200 mL of water. He has some identical metal balls. He drops three balls into the water, and they sink. The water level rises to 275mL. Shaun drops five more balls into the water. What is the new water level?

A 225 mL
B 475 mL
C 375 mL
D 400 mL
E 425 mL

110. Shaun has a 0.5 litre measuring jug. It contains 100 mL of water. He has some identical metal balls. He drops two balls into the water, and they sink. The water level rises to 200mL. Shaun drops five more balls into the water. What is the new water level?

A 325 mL
B 450 mL
C 375 mL
D 400 mL
E 425 mL

111. Shaun has a 0.5 litre measuring jug. It contains 100 ml of water. He has some identical metal balls. He drops two balls into the water, and they sink. The water level rises to 200ml. Then Shaun replaced the metal balls with cubes. Now when he drops 2 cubes to the 100 ml of water the water level rises to 150 ml. What would be the water level when he drops 3 metal balls and 3 cubes.

A 325 mL

B 450 mL

C 375 mL

D 400 mL

E 425 mL

112. The school has a target to collect $500 each day by selling tickets. The graph shows how many tickets with different values they have sold so far today.

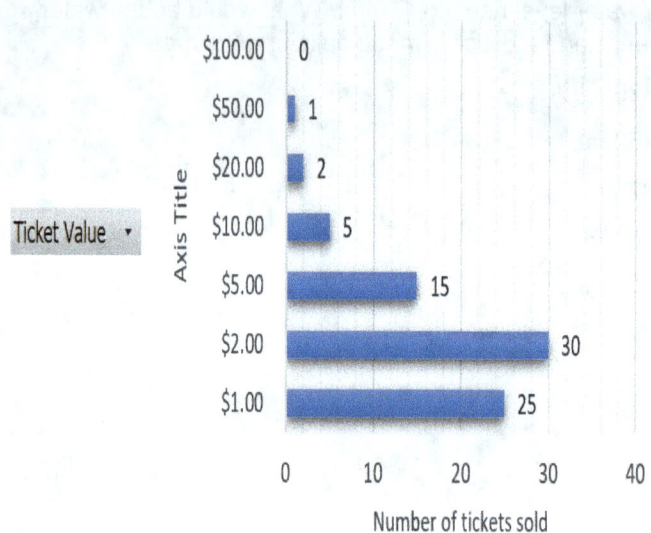

How much more money do they have to collect today?

A $100

B $200

C $300

D $150

E $50

113. The school has a target to collect $500 each day by selling tickets. The graph shows how many tickets with different values they have sold so far today.

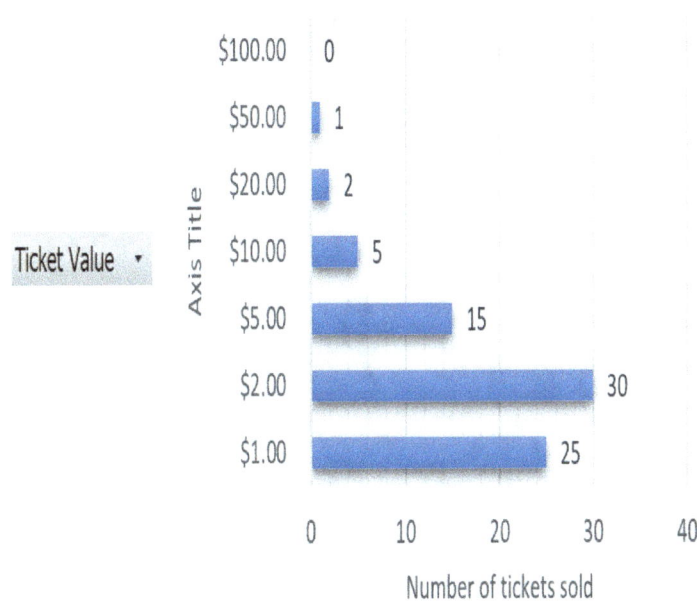

Which of the below is correct about the number of more tickets they have to sell today?

A 20 tickets of $10

B 100 tickets of $1

C 3 tickets of $100

D 5 Tickets of $50

E 200 Tickets of $2

114. The school has a target to collect $600 each day by selling tickets. The graph shows how many tickets with different values they have sold so far today.

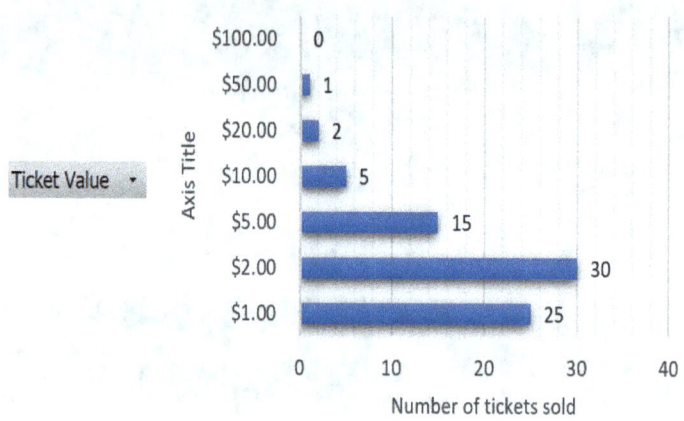

Which of the below is correct about the number of more tickets they have to sell today?

A 6 tickets of $50

B 100 tickets of $10

C 2 tickets of $100

D 5 Tickets of $50

E 200 Tickets of $2

115. The school has a target to collect $750 each day by selling tickets. Tickets are priced at $5,10,20,50 and 100.

What is the minimum number of tickets school has to sell to reach the target?

A 7

B 8

C 15

D 10

E 6

116. Kenny starts with the number 73 and follows this instruction six times:

If the number you have is odd, add 1, or if it is even, divide by 2.

After following the instruction six times, what number does Kenny have?

A 4

B 7

C 10

D 15

E 20

117. Kenny starts with the number 74 and follows this instruction six times:

If the number you have is odd, subtract 1, or if it is even, divide by 2.

After following the instruction six times, what number does Kenny have?

A 4
B 5
C 10
D 15
E 20

118. Kenny starts with the number 13 and follows this instruction six times: If the number you have is odd, subtract 1, or if it is even, multiply by 2. After following the instruction six times, what number does Kenny have?

A 384
B 192
C 96
D 48
E 12

119. If shape 1 has an area of 4 cm²

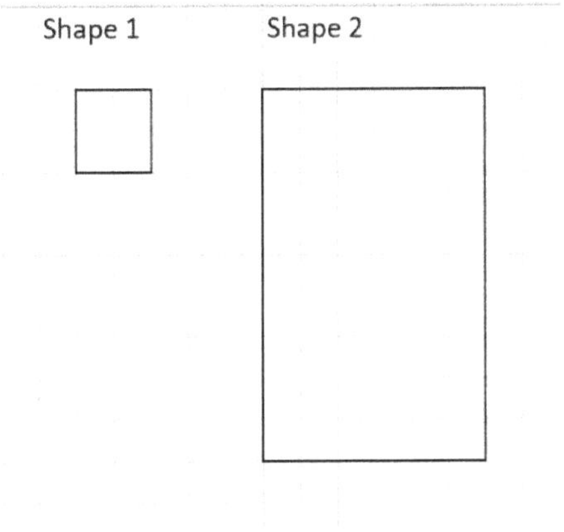

| Shape 1 | Shape 2 |

What would be the area of Shape 2?

A 54 cm²

B 50 cm²

C 65 cm²

D 46 cm²

E 64 cm²

120. Two pears weigh the same as three oranges. A shopkeeper puts fruit into bags.

Bag 1: 4 pears and 3 oranges

Bag 2: 2 pears and 6 oranges

Bag 3: 6 pears

Bag 4: 9 Oranges

Which one of these statements is correct?

A All Four bags weigh the same.

B Bags 1 and 2 weigh the same, and bag 3 and 4 have a different weight.

C Bags 1 and 3 weigh the same, and bag 2 and 4 have a different weight.

D Bags 2 and 3 weigh the same, and bag 1 and 4 have a different weight.

121. Which of these statements is/are correct about a quadrilateral?

1 Quadrilateral has 4 sides

2 Quadrilateral has 4 vertices

3 Quadrilateral has 4 interior angles

4. Trapezium is a quadrilateral.

5. Kite is a quadrilateral

A statement 1 only

B statement 2 only

C statement 3 only

D statements 4 and 5 only

E All the statements are correct

122. Which of these statements is/are correct?

1 Kite is a quadrilateral

2 Trapezoid is a quadrilateral

3 Trapezium is a quadrilateral

4. Rhombus is a quadrilateral

5 Square

6 Rectangle

A statement 4 and 6 only

B statements 1 and 2 only

C statement 3 only

D statements 4 only

E All the statements are correct

123. Little boy Kemon makes fences around square pieces of land. Each corner of the land has 1 post each. Then he puts posts 2 m apart. If each side of the land is 10m how many posts does he need in total to finish the fence around the land

A 20

B 80

C 10

D 40

E 8

Opportunity class style test-Revision Part-1

124. Little boy Kemon makes fences around square pieces of land. Each corner of the land has 1 post each. Then he puts posts 3 m apart. If each side of the land is 12m how many posts does he need in total to finish the fence around the land

A 20

B 16

C 10

D 40

E 8

125. Little boy Kemon makes fences around square pieces of land. Each corner of the land has 1 post each. Then he puts posts 3 m apart. If each side of the land is 15m how many posts does he need in total to finish the fence around the land

A 20

B 16

C 10

D 40

E 8

126. Little boy Kemon makes fences around square pieces of land. Each corner of the land has 1 post each. Then he puts posts 0.5 m apart. If each side of the land is 15m how many posts does he need in total to finish the fence around the land

A 120

B 00

C 10

D 40

E 80

127. Little boy Kemon makes fences around square pieces of land. Each corner of the land has 1 post each. Then he puts posts 0.25 m apart. If each side of the land is 15m how many posts does he need in total to finish the fence around the land

A 240

B 60

C 120

D 40

E 80

128. Little boy Kemon makes fences around square pieces of land. Each corner of the land has 1 post each. Then he puts posts 0.75 m apart. If each side of the land is 15m how many posts does he need in total to finish the fence around the land

A 240

B 60

C 120

D 40

E 80

129. These thermometers show the temperature outside and inside

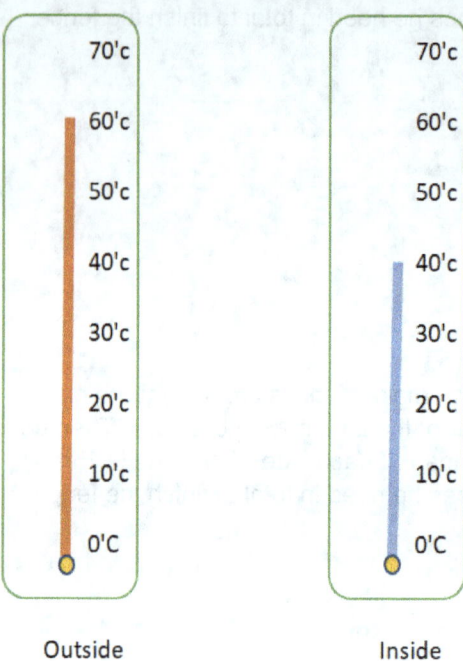

Outside   Inside

What is the difference between the temperature outside and the temperature inside?

A 20°C
B 30°C
C 40 °C
D 10°C
E 15 °C

130. Jake has three different building blocks. [diagram not to scale] He stacks all three on top of each other in different ways. What is the tallest stack he can make?

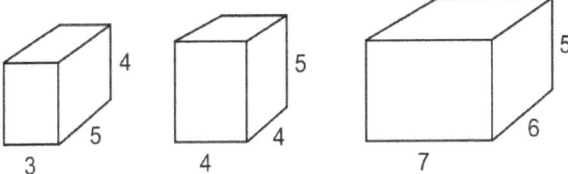

A 13 cm

B 14 cm

C 15 cm

D 17 cm

E 18 cm

131. Jake has three different building blocks. [diagram not to scale] He stacks all three on top of each other in different ways. What is the shortest stack he can make?

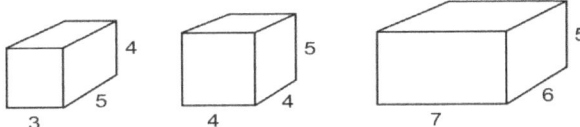

A 12 cm

B 14 cm

C 15 cm

D 17 cm

E 18 cm

132. X Box Max has two identical squares, X and Y. He divides square X into 36 equal-sized small squares.

He divides square Y into 9 equal-sized squares.

He shades some squares on X and some squares on Y, as shown

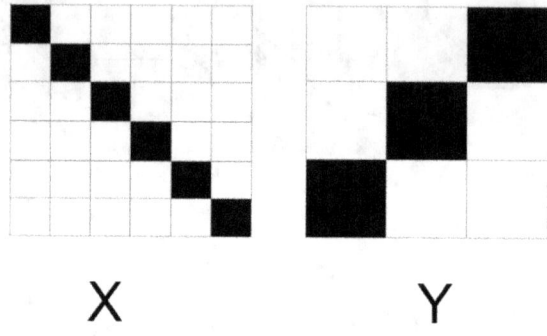

X   Y

How many more small squares must X Box Max shade on X so that the same area is shaded on both diagrams?

A 3

B 6

C 12

D 24

E 30

133. Yetty has two identical circles, X and Y. He divides circle X into 4 equal-sized pieces.

He divides circle Y into 16 equal-sized pieces.

He shades some pieces on X and some squares on Y, as shown

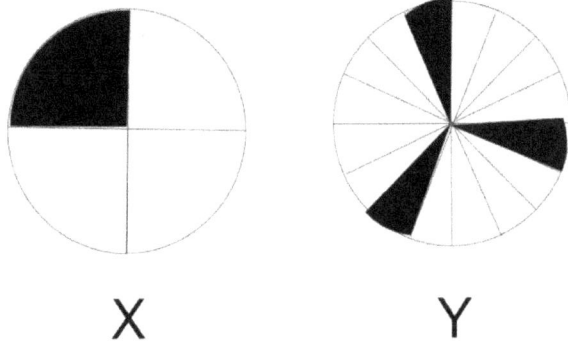

X   Y

How many more pieces must Yetty shade on Y so that the same area is shaded on both diagrams?

A 1

B 2

C 3

D 4

E 5

134. Yetty has two shapes X and Y with same area. He divides triangle X into 9 equal-sized pieces.

He divides hexagon Y into 6 equal-sized pieces.

He shades some pieces on X and some squares on Y, as shown

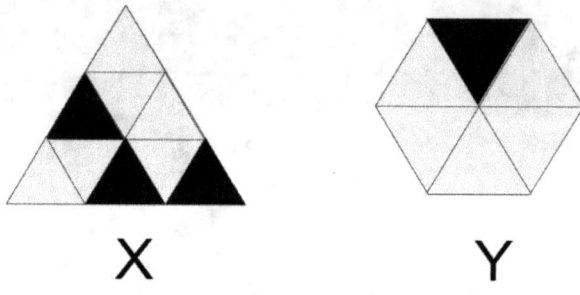

X   Y

How many more pieces must Yetty shade on Y so that the same area is shaded on both diagrams?

A 1

B 2

C 3

D 4

E 5

135. Nina has one each of the shapes below, which she can move and rotate

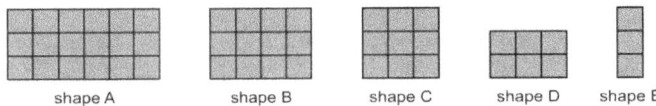

She has an empty 6 by 6 grid.

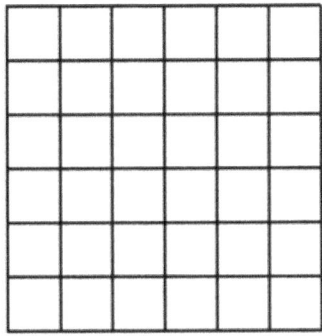

Nina realises she can cover the grid exactly using four of these shapes, with no shapes overlapping. Which shape is not used?

A shape A

B shape B

C shape C

D shape D

E shape E

136 Pamela has one each of the shapes below, which she can move and rotate

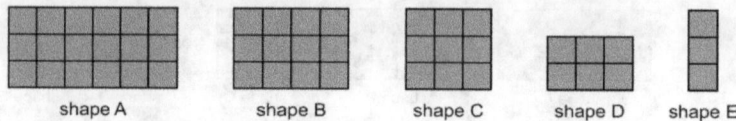

She has an empty 6 by 6 grid.

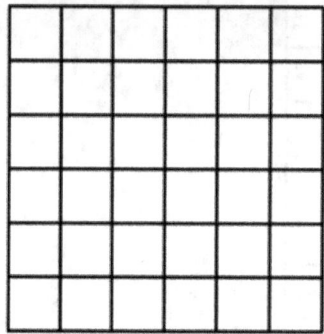

Pamela realises she can cover the grid exactly using three of these shapes, with no shapes overlapping. Which shapes are not used?

A shape A and E

B shape B and E

C shape C and E

D shape D and E

E shape A and B

137. The sum of the seven number cards below is 35.

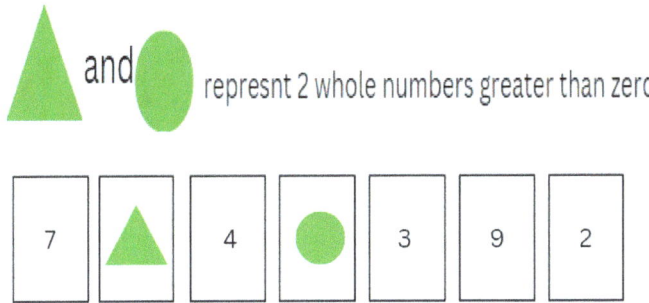 represnt 2 whole numbers greater than zero

| 7 | ▲ | 4 | ● | 3 | 9 | 2 |

What is the largest possible difference between Triangle and Circle?

A 8

B 2

C 3

D 4

E 6

138. Wen is measuring water in a container. This container has an unusual scale.

What is the volume of water Wen having?

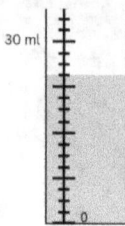

A 24 mL

B 25 mL

C 26 mL

D 28 mL

E 32 mL

139. Kelly follows these instructions:

• Choose a starting number.

• Add 4 to your number.

• Multiply by 2. He gets the answer 40.

What do you get if you add together the digits in his starting number?

A 6

B 7

C 9

D 16

E 15

140. The children in a class are making up a new type of money.

They decide that:
- 9 squigs are worth 2 zots
- 3 zots are worth 6 grozzles

How many squigs are 24 grozzles worth?

A 24 squigs

B 36 squigs

C 54 squigs

D 108 squigs

E 150 squigs

141. Seven boxes are going to be packed into bags.

Each box has a mass of 225 grams.

Each bag can hold a mass of up to 1.5 kilograms.

What is the smallest number of bags needed to pack all the boxes?

A 3

B 2

C 6

D 7

E 10

Opportunity class style test-Revision Part-1

142. Fifty tennis balls are going to be packed into bags.

Each ball has a mass of 100 grams.

Each bag can hold a mass of up to 2Kg.

What is the smallest number of bags needed to pack all the boxes?

A 6

B 5

C 3

D 7

E 2

143. Identical tennis balls are going to be packed into bags. Each ball weighs 100 grams.

Each bag can hold a mass of up to 2 Kilo grams.

What is the highest number of balls that can be packed in 3 bags?

A 3

B 50

C 60

D 6

E 5

Opportunity class style test-Revision Part-1

144. A passenger lift can carry people up to 500 kgs. If each passenger can weigh 75 kg.

What is the highest number of people can fit in the lift at one time?

A 3

B 7

C 8

D 6

E 5

145. A passenger lift can carry people up to 550 kgs. If each passenger can weigh 75 kg. When the lift is overweight, it starts alarming.

when all the passengers got in the lift the started alarming. There are 8 passengers in the lift. What is the minimum number of passengers that need to get off for the lift to stop alarming.

A 3

B 2

C 1

D 4

E 5

146. A passenger lift can carry up to 15 passengers if each passenger weights 75 kg. if each passenger in the lift weights 105kg hat is the maximum number of passengers can be in the lift.

A 9

B 10

C 6

D 4

E 5

147. A passenger lift can carry up to 15 passengers if each passenger weights 75 kg. if the lift is overweight the alarm beeps. One day the lift was in full capacity. At one stop 4 passengers off loaded the lift and a 10 identical metal objects were placed. The alarm started beeping. Once 4 metal objects were taken out the alarm stopped. What was the maximum weight the metal objects can be?

A 75

B 100

C 50

D 25

E 125

148. A passenger jet can carry up to 1500 kgs. If each passenger carries 30 Kg of baggage what is the maximum weight of each passenger to be able to carry 15 passengers?

A 70

B 100

C 50

D 25

E 125

149. A cargo ship can carry up to 15000 kgs. The ship is in full capacity of 15000kgs with identical containers weighing 25Kgs each. What is the number of containers in the ship.

A 700

B 600

C 500

D 250

E 125

150. The column graph shows the number of car tyres replaced by two mechanics, Billil and Cathy, in one day. The scale is missing from the graph.

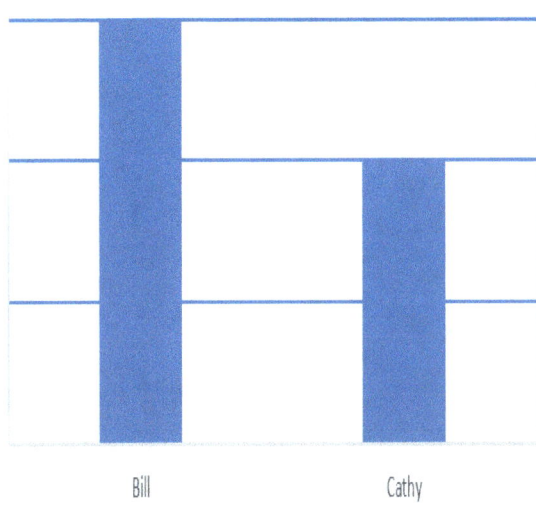

If Bill replaced 12 tyres on that day how many tyres did Cathy replace?

A 8

B 6

C 9

D 12

E 4

Opportunity class style test-Revision Part-1

151. The picture graph shows the number of apples picked up by Bill. The pictures for Cathy are missing. Both picked 1400 apples in total.

Worker	Number of Apples picked		Guide	
Bill	🍎 🍎 🍎 🍎 🍎̧		🍎	200 apples
Cathy			🍎̧	100 apples

Which of the below is correct for Cathy?

A.

Worker	Number of Apples picked
Cathy	🍎 🍎 🍎 🍎 🍎̧

B.

Worker	Number of Apples picked
Cathy	🍎 🍎 🍎 🍎

C.

Worker	Number of Apples picked
Cathy	🍎 🍎 🍎

D.

Worker	Number of Apples picked
Cathy	🍎 🍎

E.

Worker	Number of Apples picked
Cathy	🍎 🍎 🍎̧

Opportunity class style test-Revision Part-1

152. Nine children are going to play in a badminton tournament. Each child will play every other child once. How many matches will there be in total?

A 15

B 18

C 21

D 30

E 36

153. Ten school kids are going to play in a badminton tournament. Each child will play every other child once. How many matches will there be in total?

A 15

B 18

C 45

D 30

E 36

154. Three school kids are going to play in a badminton tournament. Each child will play every other child once. How many matches will there be in total?

A 3

B 6

C 4

D 5

E 7

155. Four school kids are going to play in a badminton tournament. Each child will play every other child once. How many matches will there be in total?

A 3

B 6

C 4

D 5

E 7

156. Five school kids are going to play in a badminton tournament. Each child will play every other child once. How many matches will there be in total?

A 3

B 6

C 10

D 5

E 7

157. Six school kids are going to play in a badminton tournament. Each child will play every other child once. How many matches will there be in total?

A 3

B 6

C 10

D 15

E 7

Opportunity class style test-Revision Part-1

158. Six school kids are going to play in a badminton tournament. Each child will play every other child once. How many matches will there be in total?

A 3

B 6

C 10

D 15

E 7

159. Seven school kids are going to play in a badminton tournament. Each child will play every other child once. How many matches will there be in total?

A 3

B 6

C 10

D 15

E 21

160. Eight school kids are going to play in a badminton tournament. Each child will play every other child once. How many matches will there be in total?

A 3

B 6

C 28

D 15

E 21

Opportunity class style test-Revision Part-1

161. Two school kids are going to play in a badminton tournament. Each child will play every other child once. How many matches will there be in total?

A 2

B 6

C 1

D 11

E 6

162. Here is part of a number line. 789 The arrow represents one jump. Starting from 9, what value will be reached on the number line after 11 of these jumps?

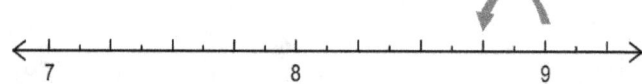

A 6 ¼

B 7 ¼

C 6 ¾

D 7 ¾

E 7 ½

163. Here is part of a number line. 789 The arrow represents one jump. Starting from 9, what value will be reached on the number line after 9 of these jumps?

A 6 ¼

B 7 ¼

C 6 ¾

D 7 ¾

E 7 ½

164. Here is part of a number line. 789 The arrow represents one jump. Starting from 9, what value will be reached on the number line after 5 of these jumps?

A 6 ¼

B 7 ¼

C 6 ¾

D 7 ¾

E 7 ½

165. Here is a bus timetable:

Forster to Coffs Harbour

Forster 7:00 am 8:50 am

Wauchope 8:48 am 10:38 am

Kempsey 9:58 am 11:48 am

Coffs Harbour 11:49 am 1:39 pm

Coffs Harbour to Forster

Coffs Harbour 2:39 pm 4:25 pm

Kempsey 4:40 pm 6:26 pm

Wauchope 5:53 pm 7:39 pm

Forster 7:33 pm 9:19 pm

Sarah and her friends live in Forster and take the 8:50 am bus to Coffs Harbour.

They stay in Coffs Harbour until they get a bus home, arriving in Forster at 9:19 pm that evening. How long do they spend at Coffs Harbour?

A 2 hours 46 minutes

B 3 hours 14 minutes

C 3 hours 26 minutes

D 4 hours 36 minutes

E 7 hours 40 minutes

166. Here is a bus timetable:

Forster to Coffs Harbour

Forster 7:00 am 8:50 am

Wauchope 8:48 am 10:38 am

Kempsey 9:58 am 11:48 am

Coffs Harbour 11:49 am 1:39 pm

Coffs Harbour to Forster

Coffs Harbour 2:39 pm 4:25 pm

Kempsey 4:40 pm 6:26 pm

Wauchope 5:53 pm 7:39 pm

Forster 7:33 pm 9:19 pm

Sarah and her friends live in Forster and take the 8:50 am bus to Coffs Harbour.

They stay in Coffs Harbour until they get a bus home, arriving in Forster at 7:33 pm that evening. How long do they spend at Coffs Harbour?

A 1 hours 0 minutes

B 3 hours 14 minutes

C 3 hours 26 minutes

D 4 hours 36 minutes

E 7 hours 40 minutes

167. Here is a bus timetable:

Forster to Coffs Harbour

Forster 7:00 am 8:50 am

Wauchope 8:48 am 10:38 am

Kempsey 9:58 am 11:48 am

Coffs Harbour 11:49 am 1:39 pm

Coffs Harbour to Forster

Coffs Harbour 2:39 pm 4:25 pm

Kempsey 4:40 pm 6:26 pm

Wauchope 5:53 pm 7:39 pm

Forster 7:33 pm 9:19 pm

Sarah and her friends live in Forster and take the 7:00 am bus to Coffs Harbour.

They stay in Coffs Harbour until they get a bus home, arriving in Forster at 9:19 pm that evening. How long do they spend at Coffs Harbour?

A 4 hours 36 minutes

B 3 hours 14 minutes

C 3 hours 26 minutes

D 4 hours 36 minutes

E 7 hours 40 minutes

168. Here is a bus timetable:

Parramatta to Caringbah

Parramatta 7:00 am 8:50 am

Burwood 8:48 am 10:38 am

Hurstville 9:58 am 11:48 am

Sutherland 11:49 am 1:39 pm

Caringbah 12:09 pm 1:59pm

Caringbah to Parramatta

Caringbah 2:39 pm 4:25 pm

Sutherland 4:40 pm 6:26 pm

Hurstville 5:53 pm 7:39 pm

Burwood 7:33 pm 9:19 pm

Parramatta 7:55 pm 9:39 pm

Sarah and her friends live in Parramatta and take the 7:00 am bus to Caringbah.

They stay in Caringbah until they get a bus home, arriving in Parramatta at 9:39 pm that evening. How long do they spend at Caringbah?

A 4 hours 16 minutes

B 3 hours 14 minutes

C 3 hours 26 minutes

D 4 hours 36 minutes

E 7 hours 40 minutes

169. Here is a bus timetable:

Parramatta to Caringbah

Parramatta 7:00 am 8:50 am

Burwood 8:48 am 10:38 am

Hurstville 9:58 am 11:48 am

Sutherland 11:49 am 1:39 pm

Caringbah 12:09 pm 1:59pm

Caringbah to Parramatta

Caringbah 2:39 pm 4:25 pm

Sutherland 4:40 pm 6:26 pm

Hurstville 5:53 pm 7:39 pm

Burwood 7:33 pm 9:19 pm

Parramatta 7:55 pm 9:39 pm

Sarah and her friends live in Parramatta and take the 7:00 am bus to Caringbah.

They stay in Caringbah until they get a bus home, arriving in Parramatta at 7:55 pm that evening. How long do they spend at Caringbah?

A 2 hours 30 minutes

B 3 hours 14 minutes

C 3 hours 26 minutes

D 4 hours 36 minutes

E 7 hours 40 minutes

Opportunity class style test-Revision Part-1

170. Ash and Beau are collecting empty pens.

Altogether they have 119 empty pens, but Ash has 9 more than Beau. How many empty pens does Ash have?

A 55

B 64

C 46

D 110

E 69

171 Karia and Chloe are collecting empty pens.

Altogether they have 128 empty pens, but Karia has 14 more than Chloe. How many empty pens does Karia have?

A 71

B 64

C 86

D 57

E 69

172. Karia and Chloe are collecting empty pens.

Altogether they have 128 empty pens, but Karia has 24 more than Chloe. How many empty pens does Karia have?

A 76

B 64

C 86

D 52

E 69

Opportunity class style test-Revision Part-1

173. Ash and Beau are collecting empty pens.

Altogether they have 77 empty pens, but Ash has 3 more than Beau. How many empty pens does Ash have?

A 55

B 37

C 16

D 40

E 69

174. Ash and Beau are collecting empty pens.

Altogether they have 17 empty pens, but Ash has 3 more than Beau. How many empty pens does Ash have?

A 5

B 6

C 10

D 4

E 6

175. Ash and Beau are collecting empty pens.

Altogether they have 96 empty pens, but Ash has 6 more than Beau. How many empty pens does Ash have?

A 51

B 60

C 10

D 45

E 46

Opportunity class style test-Revision Part-1

176. Ash and Beau are collecting empty pens.

Altogether they have 116 empty pens, but Ash has 6 more than Beau. How many empty pens does Ash have?

A 61

B 60

C 50

D 45

E 46

177. Ash and Beau are collecting empty pens.

Altogether they have 86 empty pens, but Ash has 16 more than Beau. How many empty pens does Ash have?

A 51

B 60

C 50

D 45

E 46

178. Ash and Beau are collecting empty pens.

Altogether they have 96 empty pens, but Ash has twice as of Beau. How many empty pens does Ash have?

A 64

B 60

C 50

D 45

E 46

179. Ash and Beau are collecting empty pens.

Altogether they have 96 empty pens, but Ash has three times as of Beau. How many empty pens does Ash have?

A 72

B 24

C 50

D 45

E 46

180. Ash and Beau are collecting empty pens.

Altogether they have 100 empty pens, When Ash gives 4 pens to Beau both have equal number of pens. How many empty pens did Ash have before 4 pens were given by Beau?

A 46

B 54

C 50

D 45

E 40

181.Ash and Beau are collecting empty pens.

Altogether they have 95 empty pens, Then Ash collects 5 more pens which makes Ash got 4 times more pens than Beau. How many empty pens did Beau have?

A 21
B 20
C 50
D 45
E 46

182.Ash and Beau are collecting empty pens.

Altogether they have 55 empty pens, When Ash collects 5 pens more Ash got 4 times more pens than Beau. How many empty pens did Beau have?

A 12
B 20
C 48
D 45
E 46

183. Ash and Beau are collecting empty pens.

Altogether they have 55 empty pens, Ash got 4 times more pens than Beau. Then Beau collected some pens now Ash got only 2 times more than Beau. How many pens did Beau collect?

A 12

B 11

C 13

D 15

E 16

184. Altogether they have some empty pens, Ash got 4 times more pens than Beau. Then Beau collected 11 pens now Ash got only 2 times more than Beau. How many pens did both have originally?

A 55

B 66

C 22

D 15

E 16

185. In a press, the same number of documents to be printed every day. Cupid Prints on a Thursday. If he starts at 6:15 am and works as quickly as he can, he finishes at 7:15 am. Mary prints on a Friday. If she starts at 6:25 am and works as quickly as she can, she finishes at 6:55 am.

They both prints on a Saturday. If they start at 6:35 am and work as quickly as they can, what time will they finish?

A 6:30 am

B 6:40 am

C 6:55 am

D 7:25 am

E 6:45 am

186. In a press, the same number of documents to be printed every day. Cupid Prints on a Thursday. If he starts at 6:15 am and works as quickly as he can, he finishes at 7:15 am.

Mary prints on a Friday. She prints two times as fast as Cupid.

They both prints on a Saturday. If they start at 6:35 am and work as quickly as they can, what time will they finish?

A 7:35 am

B 8:45 am

C 7:15 am

D 7:25 am

E 6:55 am

187. In a press, the same number of documents to be printed every day. Cupid Prints on a Thursday. If he starts at 6:15 am and works as quickly as he can, he finishes at 6:55 am.

Mary prints on a Friday. She prints four times as fast as Cupid.

They both prints on a Saturday. If they start at 6:35 am and work as quickly as they can, How much time will be taken to finish the job?

A 12 minutes

B 10 minutes

C 15 minutes

D 8 minutes

E 6 minutes

188. Kemon saves money by depositing in the bank account. At the start of Monday, the account has $20. At the end of Monday, the account balance is $32. At the end of Wednesday, the account balance is $50. At the end of Friday, the account balance is $100. He deposits in the same amount of money on each of Tuesday, Wednesday and Friday. How much money does she deposit on Thursday?

A $25

B $29

C $32

D $39

E $41

189. Kemon saves money by depositing in the bank account. At the start of Monday, the account has $0. At the end of Monday, the account balance is $22. At the end of Wednesday, the account balance is $50. At the end of Friday, the account balance is $100. He deposits in the same amount of money on each of Tuesday, Wednesday and Friday. How much money does she deposit on Thursday?

A $25

B $29

C $36

D $39

E $41

190. Kemon saves money by depositing in the bank account. At the start of Monday, the account has $0. From Monday to Friday he doubles the deposit every day, At the end of Friday, the account balance is $310. How much money does she deposit on Monday?

A $10

B $21

C $25

D $31

E $100

191. Kemon saves money by depositing in the bank account. At the start of Monday, the account has $0. From Monday to Friday he halves the deposit every day, At the end of Friday, the account balance is $124. What is the approximate amount of how much money she deposited on Monday?

A $64

B $32

C $16

D $8

E $4

192. Kemon saves money by depositing in the bank account. At the start of Monday, the account has $0. From Monday to Friday he quarters the deposit every day, At the end of Friday, the account balance is $341. What is the approximate amount of how much money she deposited on Monday?

A $256

B $64

C $16

D $4

E $1

193. The diagram shows a 3-sided shape.

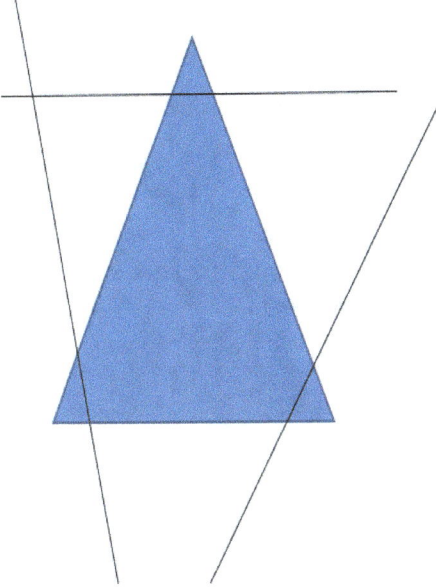

The triangle is cut by three lines shown. This makes four pieces: three triangles and one other shape. How many sides do the four pieces have in total?

   A 15

   B 13

   C 14

   D 16

   E 17

194. The diagram shows a 5 sided shape.

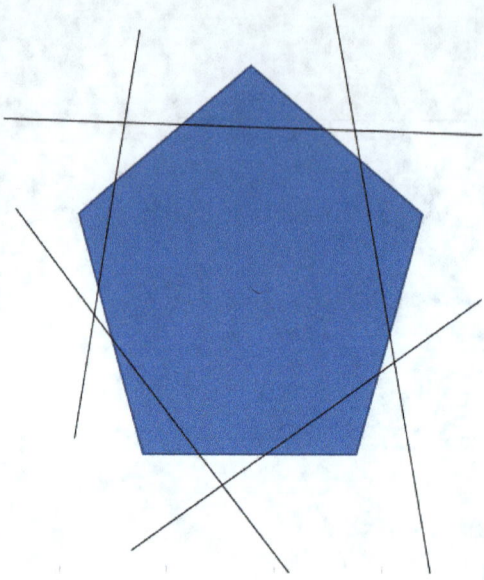

The pentagon is cut by five lines shown. This makes six pieces: Five triangles and one other shape. How many sides do the six pieces have in total?

A 15

B 25

C 10

D 16

E 17

195. The diagram shows a shape.

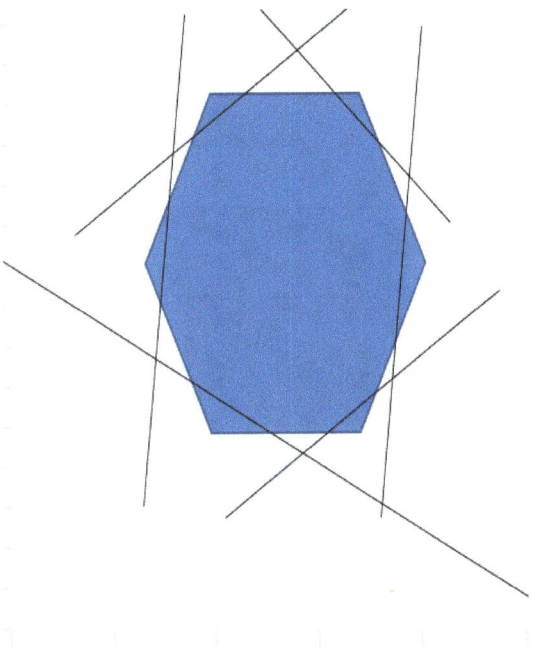

The hexagon is cut by six lines shown. This makes seven pieces: Six triangles and one other shape. How many sides do the seven pieces have in total?

A 15

B 25

C 30

D 18

E 12

196. The diagram shows 4-sided shape.

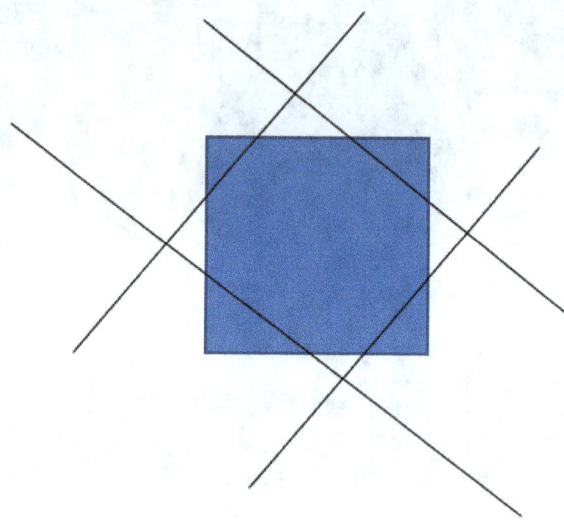

The square is cut by four lines shown. This makes five pieces: Four triangles and one other shape. How many sides do the five pieces have in total?

A 15

B 25

C 20

D 16

E 17

197. The diagram shows a shape.

This is cut by five lines shown. This makes six pieces: Five triangles and one other shape. How many sides do the six pieces have in total?

A 30

B 25

C 20

D 16

E 17

198. The diagram shows a shape.

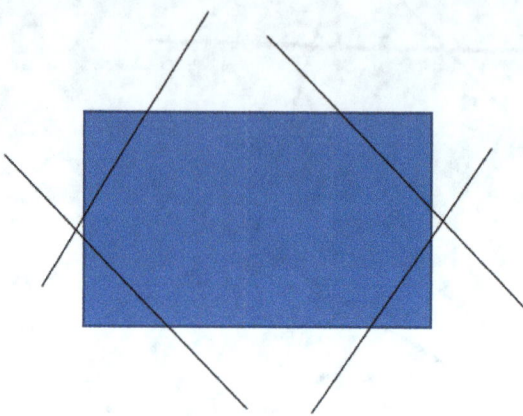

The rectangle is cut by four lines shown. This makes five pieces: Four triangles and one other shape. How many sides do the five pieces have in total?

A 15

B 25

C 20

D 12

E 17

199. The diagram shows a shape.

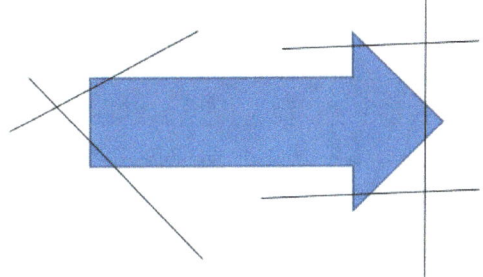

The arrow is cut by five lines shown. This makes six pieces: Five triangles and one other shape. How many sides do the five pieces have in total?

A 15

B 27

C 20

D 12

E 17

200. The diagram shows a map.

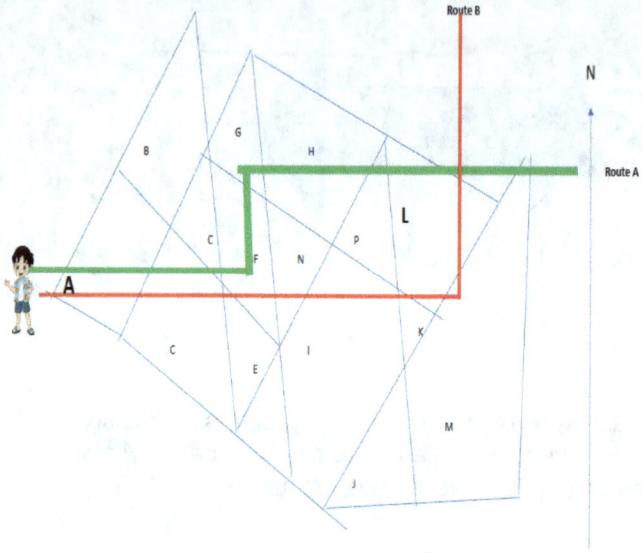

In the above picture there are different lands marked from A to M. Boy plans a route from A to L. On his routes he can only travels north or east. He can make as many turns as she wants, but always travels north or east. What is the greatest number of different lands he could pass through? (Do not include A or L.)

A 3

B 5

C 6

D 7

E 8

201. These thermometers show the temperature outside and inside

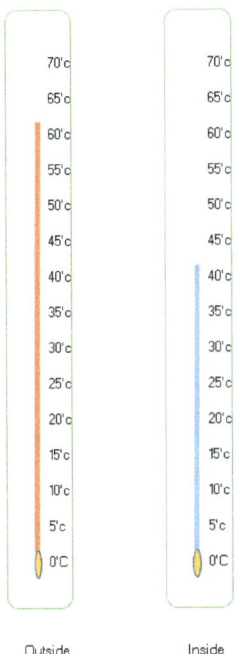

Outside   Inside

What is the difference between the temperature outside and the temperature inside?

A 20°C

B 30°C

C 40 °C

D 10°C

E 15 °C

202. Jake has three different building blocks. [diagram not to scale] He stacks all three on top of each other in different ways. What is the shortest stack he can make?

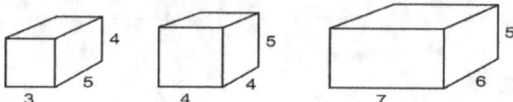

A 13 cm

B 12 cm

C 15 cm

D 17 cm

E 18 cm

203. Jake has three different building blocks. [diagram not to scale] He stacks all three on top of each other in different ways. What is the difference between the shortest stack and the tallest he can make?

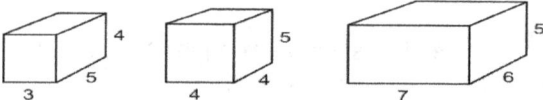

A 5 cm

B 4 cm

C 10 cm

D 17 cm

E 12 cm

204. Max has two identical squares, X and Y. He divides square X into 36 equal-sized small squares.

He divides square Y into 9 equal-sized squares.

He shades some squares on X and some squares on Y, as shown

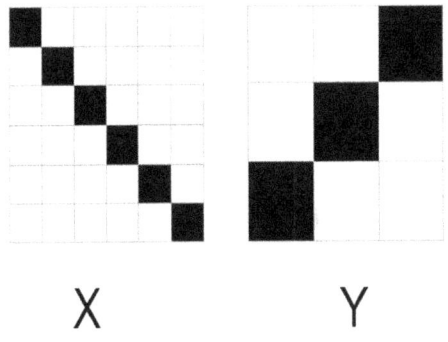

X   Y

How many more small squares must Max shade on X so that the 2 times more area is shaded in X diagram in comparison to Y diagram?

A 18

B 6

C 12

D 24

E 30

205. X Box Max has two identical squares, X and Y. He divides square X into 36 equal-sized small squares.

He divides square Y into 9 equal-sized squares.

He shades some squares on X and some squares on Y, as shown

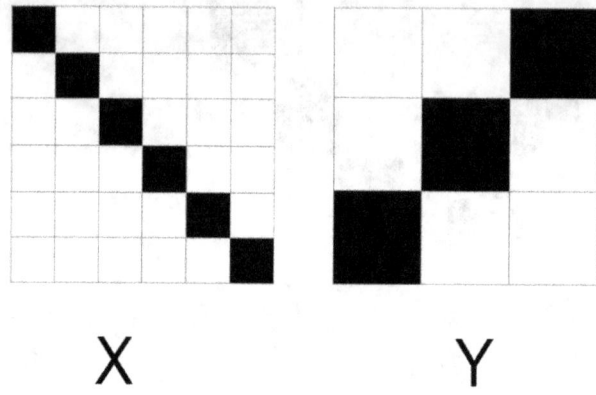

X   Y

What is the coloured fraction in X and Y diagrams

A ⅙ in X and ⅓ in Y

B ⅙ in X and ⅙ in Y

C ⅓ in X and ⅙ in Y

D ⅓ in X and ⅔ in Y

E ⅙ in X and ⅙ in Y

206. X Box Max has two identical squares, X and Y. He divides square X into 36 equal-sized small squares.

He divides square Y into 9 equal-sized squares.

He shades some squares on X and some squares on Y, as shown

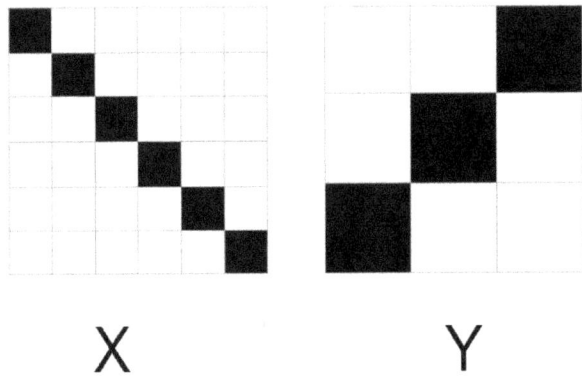

Then he decided to shade 6 more squares in diagram X. Which of the below is correct about the final diagram

What is the coloured fraction in X and Y diagrams

A ⅙ in X and ⅓ in Y

B ⅓ in X and ⅓ in Y

C ⅓ in X and ⅙ in Y

D ⅓ in X and ⅔ in Y

E ⅙ in X and ⅙ in Y

207. Bu Bu Bear wants to start a trip from Western Australia (WA) to New South Wales & ACT(NSW). He can travel only in the direction of South and East,

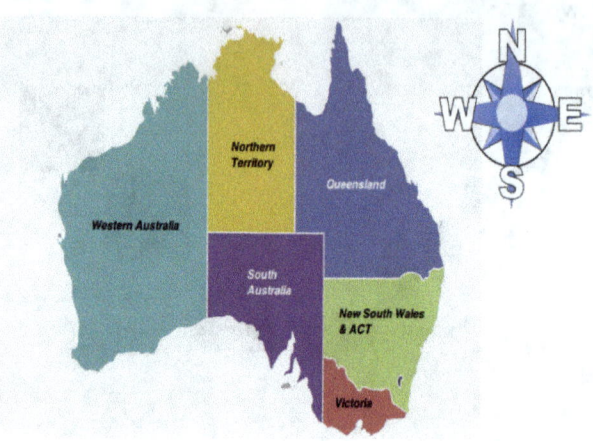

What is the largest number of states he can pass through except WA and NSW?

A 1

B 3

C 2

D 4

E 5

208. Bu Bu Bear wants to start a trip from Western Australia (WA) to New South Wales & ACT(NSW). He can travel only in the direction of South and East,

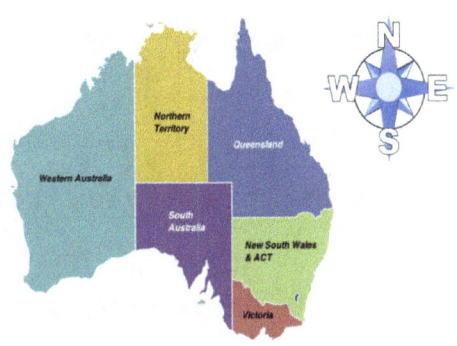

What is the smallest number of states he can pass through except WA and NSW?

A 1

B 3

C 2

D 4

E 5

209. The sum of the eight number cards below is 35. All cards have different numbers.

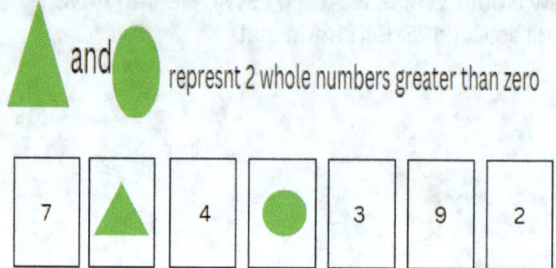

▲ and ● represnt 2 whole numbers greater than zero

| 7 | ▲ | 4 | ● | 3 | 9 | 2 |

What is smallest possible difference between Triangle and Circle?

A 0

B 2

C 3

D 4

E 6

Opportunity class style test-Revision Part-1

210. Wen is measuring water in a container. This container has an unusual scale.

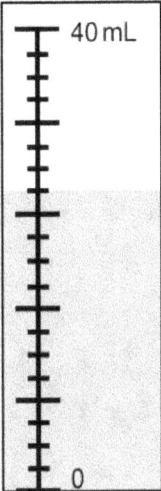

What is the volume of water Wen having?

A 24 mL

B 26 mL

C 30 mL

D 28 mL

E 32 mL

211. Kelly follows these instructions:
- Choose a starting number.
- Add 8 to your number.
- Multiply by 2.
- Divide by 3

He gets the answer 20.

What do you get if you add together the digits in his starting number?

A 22

B 7

C 9

D 16

E 15

212. Kelly follows these instructions:
- Choose a starting number.
- Add 6 to your number.
- Multiply by 2.
- Divide by 3
- Subtract 2

He gets the answer 20.

What do you get if you add together the digits in his starting number?

A 22

B 27

C 9

D 16

E 15

213. Twelve boxes are going to be packed into bags.

Each box has a mass of 525 grams.

Each bag can hold a mass of up to 1.5 kilograms.

What is the smallest number of bags needed to pack all the boxes?

A 3

B 5

C 6

D 7

E 10

214. Which of these statements is/are correct?

1 All rhombuses are parallelograms.

2 All squares are parallelograms

3 All rectangles are parallelograms

4. All rhomboids are parallelograms

A statement 1 only

B statement 2 only

C statement 3 only

D statements 4 only

E All the statements are correct

Opportunity class style test-Revision Part-1

Answers

1. C
2. E
3. B
4. E
5. C
6. C
7. C
8. B
9. E
10. C
11. B
12. D
13. C
14. C
15. B
16. D
17. C
18. C
19. B
20. D
21. C
22. B
23. A
24. C
25. B
26. C
27. B
28. D
29. D
30. C
31. C
32. B
33. D
34. B
35. C
36. B
37. A
38. B

39. B
40. C
41. A
42. B
43. C
44. C
45. D
46. E
47. E
48. E
49. E
50. E
51. E
52. E
53. A
54. B
55. D
56. A
57. B
58. A
59. C
60. C
61. D
62. D
63. A
64. B
65. A
66. D
67. E
68. C
69. E
70. B
71. E
72. B
73. A
74. C
75. D
76. D
77. E

78. E
79. E
80. E
81. E
82. A
83. B
84. A
85. A
86. A
87. A
88. A
89. A
90. A
91. A
92. A
93. A
94. A
95. B
96. A
97. C
98. B
99. B
100. A
101. C
102. C
103. A
104. E
105. D
106. D
107. D
108. A
109. D
110. B
111. A
112. B
113. A
114. A
115. B
116. C

Opportunity class style test-Revision Part-1

117. B
118. A
119. A
120. A
121. E
122. E
123. A
124. B
125. A
126. A
127. A
128. E
129. A
130. D
131. A
132. B
133. A
134. A
135. B
136. D
137. C
138. A
139. B
140. C
141. B
142. C
143. C
144. D
145. C
146. B
147. C
148. A
149. B
150. A
151. E
152. E
153. C
154. A
155. B

156. C
157. D
158. D
159. E
160. C
161. C
162. A
163. C
164. D
165. A
166. A
167. A
168. A
169. A
170. B
171. A
172. A
173. D
174. C
175. A
176. A
177. A
178. A
179. A
180. B
181. B
182. A
183. B
184. A
185. C
186. E
187. D
188. E
189. C
190. A
191. A
192. A
193. A
194. B

Opportunity class style test-Revision Part-1

195. C
196. C
197. A
198. C
199. B
200. D
201. A
202. B
203. A
204. A
205. A
206. B
207. B
208. A
209. B
210. B
211. A
212. B
213. B
214. E

www.ingramcontent.com/pod-product-compliance
Lightning Source LLC
Chambersburg PA
CBHW071832210526
45479CB00001B/103